EUROPEAN WORKSHOPS ON ECO PRODUCTS

EVALUATION OF THE "DESIGN FOR HEALTH PROFILER"

EF/96/31/EN

ABOUT THE AUTHOR

The 'Design for Health Profiler' has been designed on behalf of the European Foundation for the Improvement of Living and Working Conditions by Dr. Yorick Benjamin M.Des RCA, FRSA, Director of 'Environmental Design for Ecological Need Ltd' (EDEN). Yorick Benjamin trained as an industrial designer at Ravensbourne College of Art & Design and the Royal College of Art in the United Kingdom. His doctorate (Brunel University, London UK) was titled 'An Investigation into Eco Design Tools'. He has been a practitioner of eco design for many years and has worked for major international companies. He currently works as the Senior Researcher for the United Nations Environment Programme - Working Group on Sustainable Product Development (UNEP-WG-SPD) in Amsterdam, The Netherlands.

ACKNOWLEDGEMENTS

A special thank you to Mark Strachan & Roland Whitehead for their help with software development. Also to the 'Workshop Chairs' J.C. van Weenen, Helmut Langer and Anders Smith for their time and valuable contributions.

EUROPEAN WORKSHOPS ON ECO PRODUCTS

EVALUATION OF THE "DESIGN FOR HEALTH PROFILER" - PROCEEDINGS

Dr. Yorick Benjamin

European Foundation
for the Improvement of Living and Working Conditions
Wyattville Road, Loughlinstown, Co. Dublin, Ireland
Tel: +353 1 204 3100 Fax: +353 1 282 6456

Cataloguing data can be found at the end of this publication

Luxembourg: Office for Official Publications of the European Communities, 1996

ISBN 92-827-7790-1

Printed in Ireland

PREFACE

These proceedings describe the events that took place during the 'European Workshops on Eco Products' held by the European Foundation for the Improvement of Living and Working Conditions, Dublin, 8-9 November 1995. Central to the 4 workshops held, was the evaluation and exploration by 67 participants of a new software tool called the 'Design for Health Profiler' (DfH Profiler). The prototype software was very well received with over 60% of the participants requesting it in writing since the workshops took place. The 'DfH Profiler' is a powerful tool that allows environmental criteria related to different user needs to be managed, evaluated and scored. In this fashion the 'Profiler' has the potential to audit or assess environmental concerns in fast screening techniques or as in-depth studies. The area of environmental auditing and assessment is fraught with difficulty and ongoing discussion. The software tool presented at the workshops is designed to bring management capabilities and constructive exchange of knowledge to these important issues.

Henrik Litske
Research Manager

CONTENTS

SUMMARY

Section 1 of these proceedings explains the background to the development of the 'Profiler' and the workshop arrangements.

The participants represented at the workshop came from government, trade unions, employers and designers from the areas of urban planning, graphics, furniture and industrial design. Section 2 explains why they were invited and the method and approach used during the workshops to gain an evaluation of the software design.

Section 3 provides an overview of common findings from the four parallel workshop sessions that were held. Further detail of the participants experiences are presented in Section 4. It contains selected data taken from 'participants questionnaires' which were completed prior to attending the workshop and at its end. Supporting the data are remarks that the participants included in their written responses. This was an efficient way of gaining a great deal of data on the 'Profiler' in the 3 hours available for workshop activities.

Section 5 presents the conclusions and makes recommendations for follow up actions.

Many lessons were learnt and the participants made a very valuable contribution to the future design of the 'Profiler'. In particular they reinforced and confirmed the need for user guidance in the form of 'Reference Profiles' such as EMAS, BS 7750, ISO 14000. In addition they highlighted that information seeking facilities should be included in the software - and were very supportive of developing Internet access enhancements as suggested by the 'Profiler' developer.

Those responsible for design education were keen to integrate the tool into 'core' education for students - to explore environment and sustainability issues in relationship to product development.

Recognition of a wide range of user groups who could benefit from the tool emerged as did its potential to facilitate the exchange of information and view points for different stake holders in the development of products.

Recommendations at the end of these proceedings propose that the 'Profiler' is tested further in different situations that lead to the development and support of more 'sustainable production and consumption'.

1. INTRODUCTION

This introduction covers the background to the workshop, basic concepts of the 'DfH Profiler' software and its requirements.

1.1 Where, When & Why

The 'European Workshop on Eco Products' was held at the 'European Foundation for the Improvement of Living and Working Conditions' in Dublin, Ireland, on the 8th and 9th November 1995.

The objectives of the workshop were to:

1. Evaluate the underlying concepts of the 'Design for Health' Profiler (a software prototype) and make recommendations.

2. Formulate recommendations for the European Foundation research programme in the area of product development and environment. (conclusions given in Section 5).

This report on the workshop proceedings focuses on the first objective only and is mainly based on the results of two questionnaires; one of which was sent out prior to the workshop and one completed after the workshop by the participants. Within the questionnaire participants were able to add their own comments. Supporting the questionnaire and the participants comments are general observations by the 'Workshop Chairs' and suggestions made at the Plenary Sessions.

A short description of the 'Profiler' software demonstrated at the workshops and its underlying concepts follow (also see Figure 1, Page 13).

1.2 Basic Concepts - 'DfH Profiler' Software

The 'DfH Profiler' software is a 'tool' to help designers deal with the implications of their design decisions that effect the environment and the health of workers. However, this is a complex area and a fuller description of the concepts behind the 'Profiler' can be found in the working paper: WP/95/09/EN by Yorick Benjamin. Please contact the European Foundation for copies.

The 'Profiler' is based upon several 'key' concepts to meet the demands of such an enormous subject area (social, economic and ecological aspects of living and working conditions in relationship to products) and different types of user:

Concept 1 • It is a Very Flexible 'Tool' to Meet Different Needs - based upon the assumption that people using the software are best able to decide on the 'topics' they need to measure - topics that probably relate to their everyday work experience. Topics are written into the 'topic' box by the user (Figure 1, page 13). It shows 'topics' that relate to the life cycle of a raw material which is turned into a product and finally disposed of. However, the software could be used to measure 'Stress in the Workplace' or how much 'Energy' the photocopier in an office consumes. The user simply changes the 'topic' box text to make the 'Profiler' suitable for their project needs - this makes it useful in hundreds of areas (in theory).

Concept 2 • It is a 'Tool' that Shows the Nature of the Information it Uses Clearly based on the assumption that it is not possible to quantify everything that you may want to measure. Some things we try to measure are quite simply 'subjective value judgments'. Therefore the 'Profiler' allows the user to show whether the result presented is factual and based upon science (quantitative), subjective and value laden (qualitative) or a combination of both (inference). (Figure 1, page 13)

Concept 3 • That People Need a 'Tool' that Can be Used for Quick, Inexpensive, Simple Studies and In-depth Exhaustive Ones based on the knowledge that some problems are more complex than others! The 'Profiler' shown at the workshop could 'manage' as few as 3 'topics' or as many as 512! All these 'topics' can be individually measured between 0-100% and are attached to 'spokes' (Figure 1, page 13).

Concept 4 • That it Should be Easy to Gain an Overview of the Main 'Topics' being Measured by even non-specialist users. The 'Profiler' meets this need by showing its results as a graphic pattern - called the 'Profile Pattern'. More detailed information is stored in the 'Notes' facility (Figure 1, page 13).

FIGURE 1. The basic features of the 'DfH Profiler'.

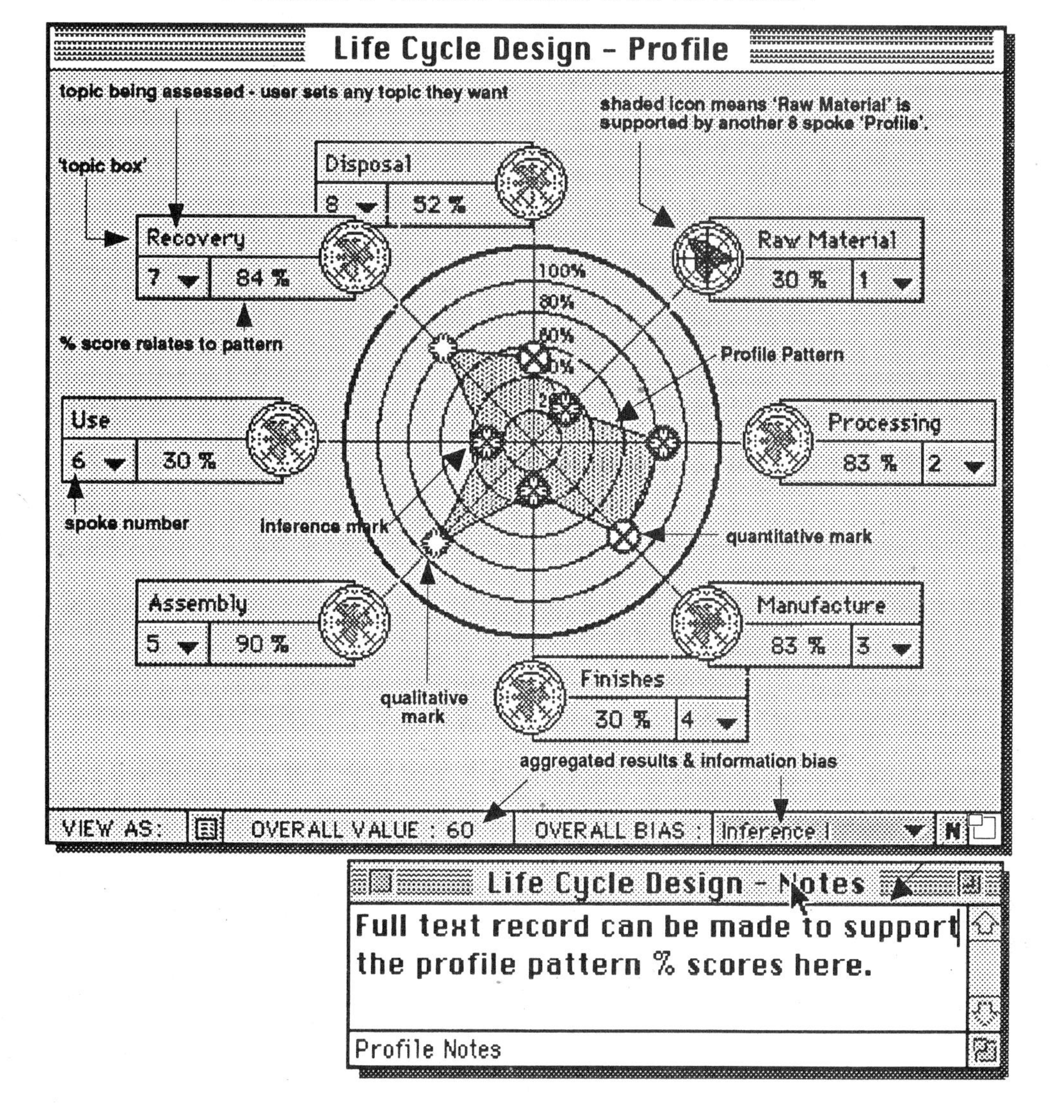

Some features of the 'Profiler' are shown above. Any 'topic' can be linked to other 'Profiles' which also contain 'topics'. The scores of all the underlying 'topics' are aggregated to support the 'topic' to which they are linked. e.g. Raw Material is supported by another 'Profile' in which all the 'topics' score 30%. The 'Profiler' can handle hundreds of topics at the same time.

1.3 Background to the 'DfH Profiler'

The 'Profiler' prototype presented at the workshops was a direct development of a previous project (by Yorick Benjamin) that concluded with the design on paper of a computer based interactive tool that had the potential to reduce environment and health impacts introduced into product design by designers. Such a 'tool' was recognised to be of importance as it is appreciated that designers have a fundamental influence in introducing impacts to products but are ill equipped to deal with the issues.

The some of the main research recommendations of the 'tool' designed on paper were that the software must be able to:

- organise, assign and weight values associated with information gained by a user
- be very transparent in how results are achieved
- allocate different information types to each weighted value shown
- manage multi-criteria issues across the whole life-cycle of processes, products and packaging
- present information in the form of a graphic profile which is supported by a written archive.

Following the research the 'Design for Health Profiler' was prototyped in software form.

The 'Design for Health Profiler' meets the need for a management tool to assist designers, policy makers, legislators, regulators and SMEs, to organise and manage the enormous wealth of information that exists on both environmental and health grounds, which is currently unmanageable, expensive to access, and organise. The 'Profiler' is a powerful interface between those who generate information and those who need to use it in multi-criteria applications.

The 'DfH' Profiler prototype software application was evaluated at the workshops by the participants; primarily in a 'questionnaire' format.

1.4 Aims of the Software Prototype

Prior to the workshops the main objectives for the DfH Profiler software were:

- to further develop the concepts presented in Phase 2 of the 'Design for Health Project'.
- to create a basic interactive software tool that encompassed 'Eco Design' and 'Design for Health' concepts, to improve the health of workers and users associated with products.
- to create a tool that would be transparent and work with multi-critera assessments.
- to realise a satisfactory prototype that could be tested by peer group review in the proposed 'European Workshops on Eco Products'.

1.5 'DfH Profiler' Prototype Features

Before the software development began the brief identified the following functions that the 'DfH Profiler' prototype would achieve:

- weighting an impact and assigning an information type with calculus
  software that creates Eco Profiles and presents percentage scores of inference, quantitative and qualitative weighting
- facilities that allow multi-criteria studies to take place
- on screen Eco Profile patterns to be generated and designed by the user
- a refined easy to use graphic interface
- the creation of an archive that supports the on screen Eco Profile

All the above were accommodated in the prototype software.

1.6 Practical Constraints & Achievements

Budget & Time

Software development is both expensive and time consuming and to create a functional software prototype within the available budget and time was a real challenge. There were genuine development constraints of budget and time that need to be highlighted as these factors effect software development and how it could be presented to the participants.

A cost constraint that arose was that it was not possible to give the participants 'hands on' experience of the software due to the expense of installing so many computer systems. Many positive requests have been noted in the 'Part 2 Questionnaires' from participants regarding the desire to use the software themselves so that they could have time to assess its usefulness in their work. In fact, since the workshops, over 60% of the participants have requested the software in writing.

Expectations

Users of software, naturally enough, have high expectations of it as they often buy inexpensive applications such as word processors that work smoothly. However, it should be recognised these programmes have taken many years and millions of dollars to develop and are by no means comparable to the 'DfH Profiler' software prototype.

Information Support & Reference Templates

There was the known requirement to provide information and references in the software system (goals, targets, etc) to support the user. However, these features were way beyond the limits of budget and time.

Achievements

The 'DfH Profiler' prototype expressed its potential to be a full application well, as the results of the questionnaires show. It was a sophisticated prototype well beyond expectations of budget and time available. It served its purpose very well in refining components of functionality, stability and user friendliness and establishing a development path. Lessons were learnt and the main recommendations are being integrated into a more complete software application.

1.7 Workshop Equipment

Each workshop coordinator was provided with an Apple Macintosh based system running an Liquid Crystal Display on an overhead projector. There were 4 systems in all.

2. METHODOLOGY & APPROACH

This section describes the design of the workshop, the background of the participants and the use of questionnaires for evaluation purposes.

2.1 Overall Approach

To evaluate the 'DfH Profiler' the following approach was organised within the resources available:

- *Multidisciplinary participants* Participants were invited who represented a wide range of professional backgrounds and countries from all over the European Union. There were 67 participants in all - not all completed questionnaires.

- *Questionnaires* Two questionnaires were provided; the first prior to attending the workshop and a second to be completed after seeing the 'DfH Profiler'.

- *Workshops* 4 workshops were run in parallel:

Graphic Design	Furniture Design	Industrial Design	Urban Planning & Sustainability
Workshop Coordinator	Workshop Coordinator	Workshop Coordinator	Workshop Coordinator
Helmut Langer	Yorick Benjamin	Anders Smith	Hans van Weenen

The time allocated to the workshop was:

Workshop Activity	Day 1	Day 2	Totals
Software Exploration	2 hours	1 hour	**3 hours**
Nº 2 Questionnaire Time	30 minutes	1 hour	**1 hour 30 minutes**
Plenary Sessions	1 hour on future research activities of the European Foundation	45 minutes on the DfH Tool workshop	**1 hour 45 minutes**

The final plenary session was for the 'workshop coordinators' to present their experiences and for participants to present their experiences.

2.2 Multidisciplinary Participants Invited

In theory the 'DfH Profiler' prototype has a very wide range users for different applications. To substantiate this theory participants from very diverse design areas were invited, namely, furniture, industrial, graphic and urban planning. In addition participants attended from trade unions and policy making bodies - they could join which workshop they wished.

Participants represented the following countries: Denmark, United Kingdom, Germany, Finland, Sweden, Austria, Greece, Ireland, Italy, The Netherlands, Belgium, Norway, France, Spain and Portugal.

Some were very familiar with computing environments whilst others were not. A minority were experienced and involved with environment and health impact issues - the majority did not have this experience. However, completed Part 1 Questionnaires told the story that without exception all participants were certain about the importance of the area.

2.3 Questionnaires

The questionnaires were designed to compliment one another; the first was sent out prior to the participants seeing the 'DfH Profiler' to gain an insight into their experience and needs in dealing with environment and design concerns. As background reading the working paper WP/95/09/EN 'The Design for Health Eco Profiler - Information Management in a Computing Environment' published by the European Foundation, was sent out with the first questionnaire.

The second questionnaire was completed after seeing the 'DfH Profiler' to gain insight on whether the 'DfH Profiler' concept could meet the participants' needs in their area of design practice. Another key objective was to gain recommendations towards improving the software. All questions followed the procedure:

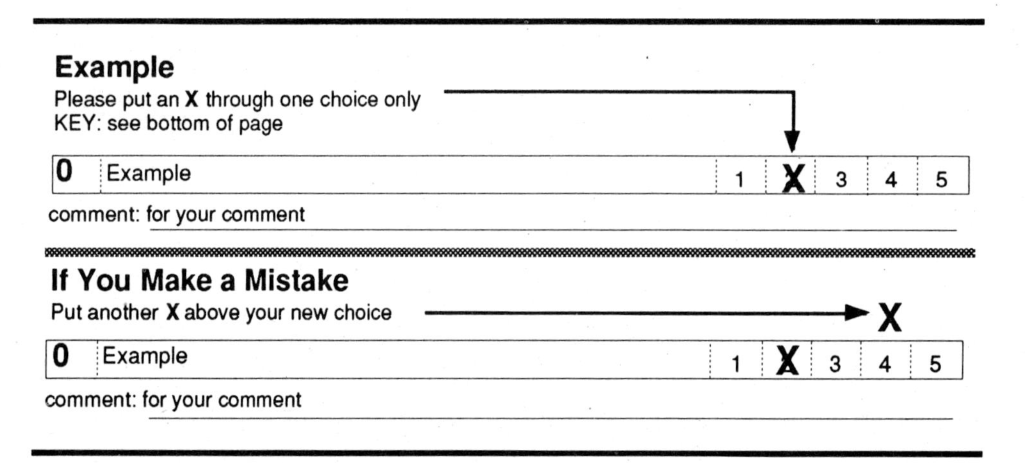

Example

Please put an **X** through one choice only

KEY: see bottom of page

0	Example	1	X	3	4	5

comment: for your comment

If You Make a Mistake

Put another **X** above your new choice → X

0	Example	1	X	3	4	5

comment: for your comment

They all used the same key:

KEY	1	2	3	4	5
	Strongly Agree	Agree	Neither Agree or Disagree	Disagree	Strongly Disagree

The following two questions are taken from Questionnaires 1 and 2 respectively. They demonstrate the comparison technique employed.

From Participants Questionnaire - Part 1 (prior to seeing the 'DfH Profiler')

3.0 ECO DESIGN QUESTIONS - OVERVIEW						
1	it is important for designers to be able to understand the environmental implications of their design decisions	1	2	3	4	5

Taken from Participants Questionnaire - Part 2 (after seeing the 'DfH Profiler')

3.0 'PROFILER' CONCEPT - OVERVIEW						
1	using the 'Profiler' concept would help me to understand the environmental implications of my design decisions	1	2	3	4	5

Limitations to the Questionnaires

Questionnaires are useful in that they focus the respondents answers to a specific issue - this is also their weakness. If the question is not asked then the interviewer will not get the answer, this leaves gaps in a study. Even when a question is asked respondents quite naturally have their own interpretation of it and it can still be difficult to gain an overview of their thinking. One method to overcome this weakness is to leave space for participants comments under each question - and this opportunity was given after all questions/statements.

In the 3 hours allotted to the software aspect of the workshop the questionnaire approach worked well. A lot of data was gained in a very short period of time in what are very complex areas; environment and software design.

2.4 Workshop Approach

The 'DfH Profiler' concept was presented, tested and explored for the first time. Due to time constraints it is important to note that the method used in the workshops was fast screening. There was not enough time to use the 'Profiler' to do comprehensive LCA like studies which are information intensive although the potential for this sort of use does exist within the design. During the workshops the 'Profiler' was used to scope, manage and give values to environmental 'topics' related to a subject selected by the workshop coordinator.

Workshop Aims

- to introduce and familiarise the participants with the DfH 'Profiler' Tool
- prioritise environmental concerns
- to fast screen a subject chosen by each coordinator
- to use the 'Profiler' in group discussions for consensus building on environmental concerns
- for the participants to agree 'values' (% scores) to environmental issues

- develop graphic 'profiles' and 'notes' on % scores given by participants.
- complete the 'Participants Questionnaire' for evaluating the Profiler concept and make recommendations.

Workshop Coordinators Role

Each workshop coordinator had the responsibility to use the 'Profiler' as they wished and to support the participants in evaluating the subject they had proposed.

Workshop coordinators could develop 'Profiles' (with the participants input) which considered a mix of objective and subjective concerns. The objective concerns leaned more towards quantification, for example: material joining techniques for disassembly, legislation, wear and tear of materials used, etc. Integrated with these concerns in the 'profile' could be the more subjective - qualitative aspects. For example: fashionable obsolescence, usability, personal preferences in colour and materials, emotional attachment, etc. This mix explored the flexibility of the 'Profiler' which can identify 3 information types: quantitative, qualitative & inference[1]. The overall approach being taken was a fast screening eco audit.

Role of Participants

The DfH software application was developed enough to demonstrate the underlying concepts and to indicate its potential. The role of the participants was to:

- evaluate the 'Profilers' underlying concepts
- contribute recommendations on the 'Profilers' future development

To stimulate meeting the above objectives the workshop coordinators supplied a subject for discussion that could be assessed or audited by the participants using the 'DfH Profiler'. Information to support the study being undertaken was to be drawn from participants by the workshop coordinators. The information management capacity of the 'Profiler' was to be explored. However, the workshops were not expected to produce detailed studies but to just explore the potential of the 'Profiler' concept. Therefore, it was not the intention that the participants should finish an audit on a product or subject - it was much more important that participants become familiar with the 'Profiler' and made a contribution to its development via discussion that the subject stimulated and by identifying strengths and weaknesses in the 'Profiler' design.

The four workshop study subjects are described in Section 3, 'Workshop Findings & Evaluation'.

1 See Benjamin. Y. 'Environment - Design & New Materials' (1991), European Foundation, & Benjamin. Y. et.al 'New Materials for Environmental Design', European Foundation (1994), ISBN 92-826-8612-4.

3. WORKSHOP FINDINGS & EVALUATION

Within the scope of these proceeding this section tries to give an overview of experiences.

3.1 Summary of 4 Workshop Experiences

Each workshop coordinator introduced a subject so that the participants could start to contribute to a study using the 'DfH Profiler'.

The subjects differed enormously between workshops, as shown below.

	Graphic Design	**Furniture Design**	**Industrial Design**	**Urban Planning & Sustainability**
Workshop Coordinator	Helmut Langer	Yorick Benjamin	Anders Smith	Hans van Weenen
Study Subject	Conference badge	Comparison of two office chairs	Disposable coffee cup filter	Make central Dublin a pedestrian area

Graphic Design Study Subject

The graphic group assessed the participants workshop badge. It was very similar to the type of badge that you would get at many international conferences. Environmentally speaking it was indefensible - a paper label with the participant's name laminated in plastic, secured to clothes by an elaborate chromium plated steel clip. It is designed as a one 'event' product as the organiser's logo is predominant on the front - otherwise participants might have been willing to use it at other events.

Furniture Design Study Subject

The group started with a modern office chair to explore using the 'Profiler'. A second chair was provided half way through the workshop as participants felt that they needed a reference product - one chair to be compared against the other. It was estimated that there was 30-40 years of age difference between the two chairs. The newer one (1995) had many non sustainable resources employed in its manufacture such as traditional oil based polymers, synthetic fibres and aluminium. It had 5 castor wheels. The older one was a simpler design in wood, steel and aluminium and made with less complex manufacturing technologies. It had 4 castor wheels.

Industrial Design Study Subject

The industrial designers were given a coffee filter to evaluate. The filter is designed so that each participant can make their own cup of coffee. It is a very irresponsible product in environmental terms although it does make a good cup of coffee! This one way trip disposable product uses valuable (environmentally speaking) finite resources: oil is exploited, refined, processed, injection moulded into the coffee filter form and a lid made. The filter is adhered to a bleached paper coffee bag, packaged, transported thousands of miles and stored at various locations. It is eventually used for 2 minutes to make a cup of coffee - then immediately thrown away, collected by the waste authority responsible and land filled or incinerated - what a waste!

Urban Planning Group

The urban planning group started by considering the 'centre' of Dublin which has a lot of through traffic and congestion. They were to use the 'Profiler' to assess the viability of making the centre a pedestrian area. They also considered a children's playground.

3.2 Common Findings

Each study subject was just a starting point for discussion. As the sessions developed participants talked more about the 'Profiler' concept, its potential and areas for improvement, and spent less time on the study subject selected by the coordinators. All groups wanted more information to support their judgments related to their study subject. The majority of participants were also keen that the 'DfH Profiler' gave them a reference template so that they could compare the subject they were presented with by their coordinators with one that had already been evaluated (a reference product). With this in mind it was suggested to the groups that they might develop their own reference templates so that others might use them as a guide in the future. *For example:* the furniture group was invited to determine the criteria for the ideal sustainable office chair. However, there were many different approaches from the group; some thought that policy was the most important criteria; others material content; some that there had to be a market for an environmental chair and therefore that the customer should contribute to the reference. None of the workshop groups took up the challenge to create a reference template themselves. Generally they thought themselves unqualified to make such a template or not informed enough on the subject.

However, discussion in all groups was productive and a valuable source of opinion and advice for future development. The talks that took place were quite appropriate as they contributed more to the development of the 'Profiler' than subject themselves - as was the underlying intention by the organisers.

There were comments about features the software needed to enhance it, however, they nearly all referred to the same two concerns:

1. a need for information to be available about the study subject
2. targets and goals required in the form of 'Reference Profiles' to support a study

Neither of these two points was achievable within the budget available.

In contrast to the opinion that the above needs had to be incorporated there was still a vast majority in favour of the 'DfH Profiler' concept and its potential as presented. As a guide to genuine enthusiasm for the 'Profiler' over 60% of the participants have written to the European Foundation requesting the software following the event.

Clarification of the participants thinking and their evaluation of the 'Profiler' was given in their written replies and questionnaire responses, extracts of which are given in the next two sections of these proceedings.

4. SAMPLES FROM THE 'EVALUATION' QUESTIONNAIRE

This section of the workshop proceedings highlights some important questionnaire responses. They are presented in the before and after comparison approach described in 'Section 2. Method & Approach'. The questions are supported with selected comments by participants from specific workshop groups. Responses must be viewed with the multidisciplinary nature of the groups in mind, the varied levels of individual knowledge on computing and environment issues - and the time available.

Please note that numbers in the bar charts relate to the number of participants answering.

4.1 GRAPHICS Design Workshop - Coordinator: *Helmut Langer*

Nº of Paired Questionnaires[2]: *8 pairs (16 questionnaires in total).*
Sample Questions from 'Participants Questionnaires: *Part 1: Sections 3 & 5 (Before workshop) & Part 2: Sections 3 & 5 (End of workshop)*

BEFORE WORKSHOP - Part 1: Section 3 - Q1 - It is important for designers to understand the environmental implications of their design decisions.

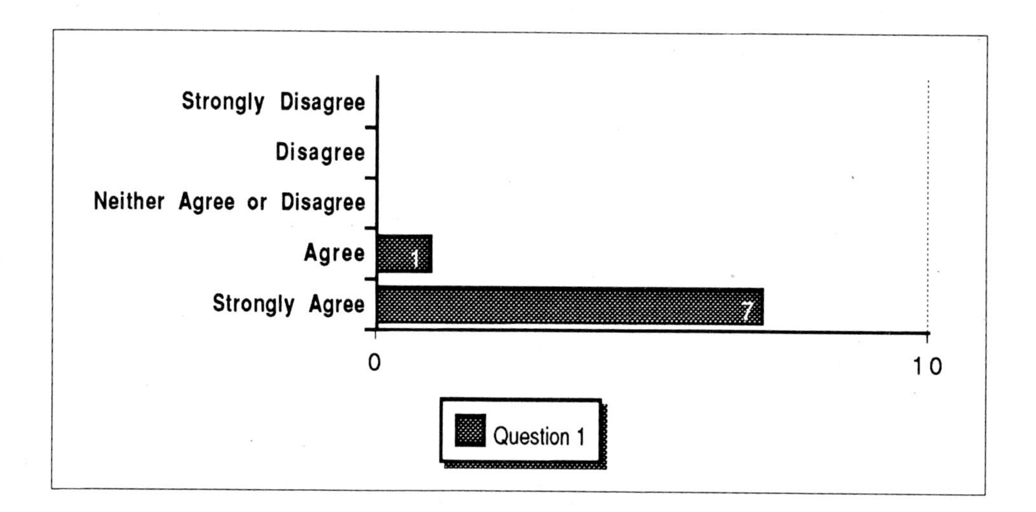

All the graphic designers agreed with the sentiment of the statement posed in Q1. Their replies were in keeping with the other 3 groups. One designer pointed out that it was important to introduce this awareness while they were studying their profession as students...*This should be applied to Design Education as well'*...and...' *Due to them (designers), being integral to finding solutions to 'non-end of pipe' environmentally benign concepts and systems'*. In other words; 'prevention' is better than 'cure' and design has a key role in 'preventing' negative environmental impacts.

---

[2] Only results have been used where the participant answered 'before' and 'end of' workshop questionnaires to make a 'pair'. All workshops had more participants than fully completed questionnaires.

WORKSHOP END - Part 2: Section 3 - Q1 - using the 'Profiler' concept would help me to understand the environmental implications of my design decisions.

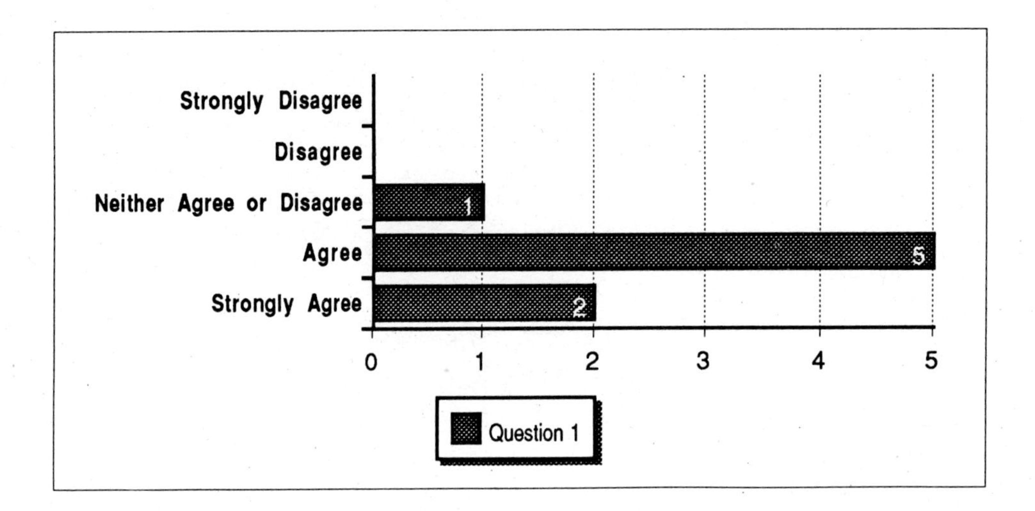

The result shows that the graphic designers felt that the 'Profiler' would be a very helpful tool. An important comment in relationship to Q1 was: *If the correct data, structuring, bench marking etc was incorporated into the programme'*. As observed in all the workshops this designer was seeking supportive information (data) and reference templates (bench marking) to be incorporated into the system. One of the questions posed connected to information was Q15.

BEFORE WORKSHOP - Part 1: Section 5 - Q15 - Evaluating environmental information is a problem because of a lack of transparency (it is not obvious where the information has come from).

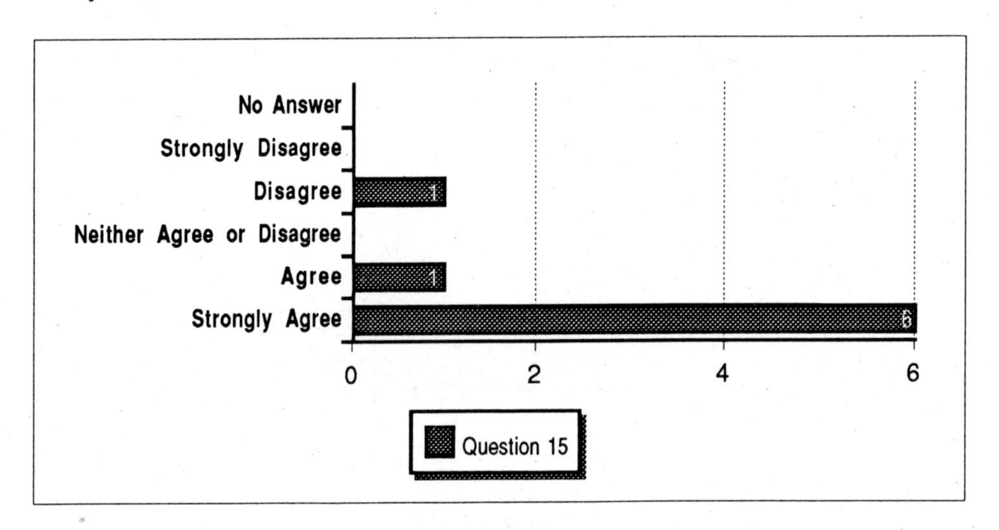

75% of the graphics group strongly agreed with the statement of Q15 and a remark was...
'Big problem! Also interpretation and evaluation of data by other parties'

<u>WORKSHOP END - Part 2: Section 5</u> - Q15 - The way the 'Profiler' concept develops its environmental 'Profile' makes the information the 'Profile' is based upon very clear.

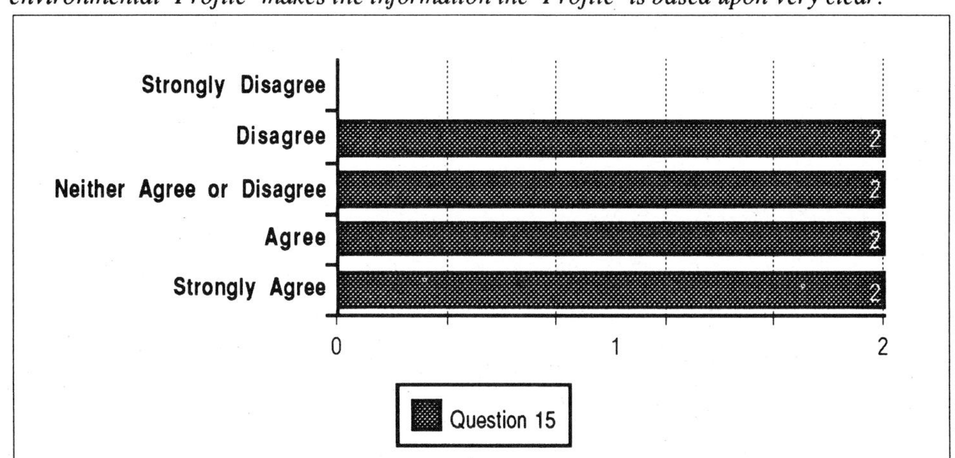

Even though the answers are spread there was still 50% agreement with this quality of the 'Profiler', whilst 25% are non decided and 25% in disagreement. These answers were given even though there was not any information available in the system and the participants themselves were responsible for developing qualitative 'Profiles'. A comment that recognised the potential and pitfalls was...'*Could be brilliant, but need to watch who collects data, how it is used and inputted*'.

4.2 FURNITURE Design Workshop - Coordinator: *Yorick Benjamin*

Nº of Paired Questionnaires: *7 pairs (14 questionnaires in total).*
Sample Questions taken from'Participants Questionnaires: *Part 1: Section 4 (Before workshop) & Part 2: Section 4 (End of workshop)*

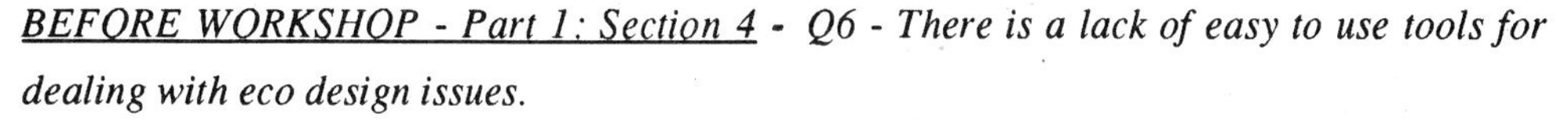

<u>BEFORE WORKSHOP - Part 1: Section 4</u> - Q6 - There is a lack of easy to use tools for dealing with eco design issues.

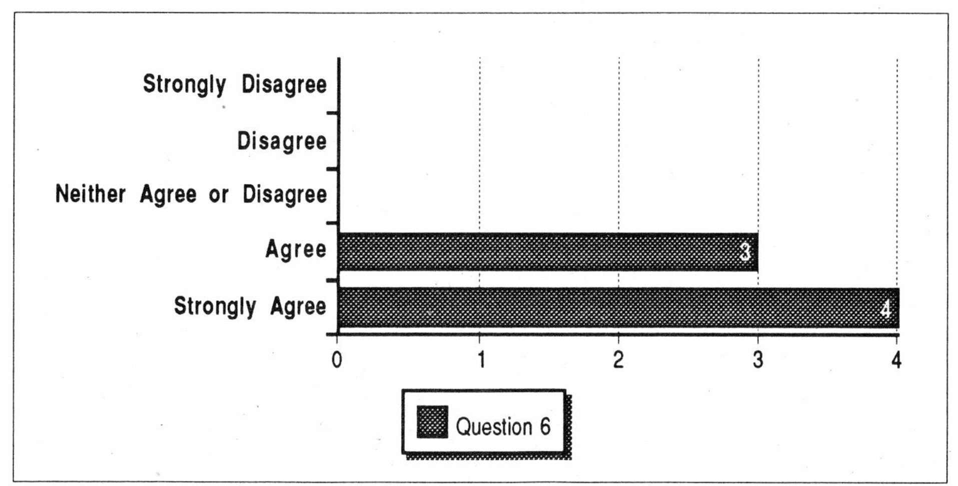

In this case all the furniture group were in agreement - this was a very similar situation to the other groups apart from Industrial Design where 16% disagreed.

WORKSHOP END - Part 2: Section 4 - *Q6 - The 'Profiler' concept was easy to use*

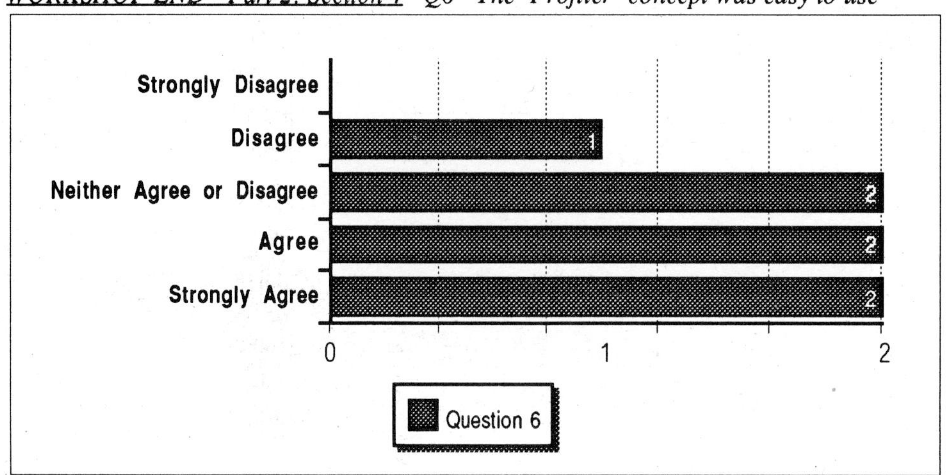

The answers are again quite spread. However it is clear that generally the view is that the 'Profiler' concept was easy to use. The furniture designers made very few 'optional comments' throughout their completed questionnaires - although this statement was made: *'Very easy to understand and to use but complicated because of definition discussions'*.
In highlighting 'definition discussions' the comment reflected again the need for reference templates to help define how, why and when to use the 'Profiler'. In the short time available of 3 hours for all workshops this was a recurring discussion point.

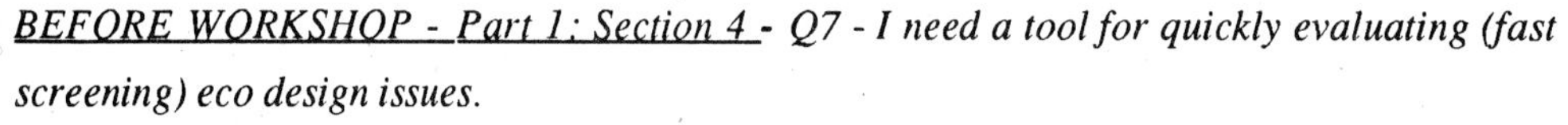

BEFORE WORKSHOP - Part 1: Section 4 - *Q7 - I need a tool for quickly evaluating (fast screening) eco design issues.*

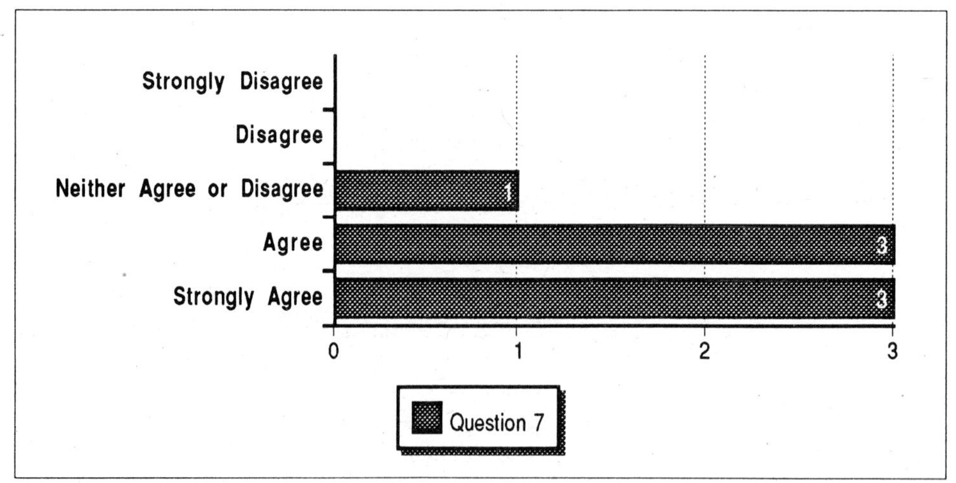

Here, the need for this type of facility is confirmed. In the design process many value judgments are traded one against another and environmental issues need to be quickly evaluated - particularly as there can be many design options available. A good fast screening tool will be more cost effective than complex LCA which can be prohibitively expensive - especially for SME's.

<u>WORKSHOP END - Part 2: Section 4</u> - Q7 - The 'Profiler' is good for fast screening environmental issues.

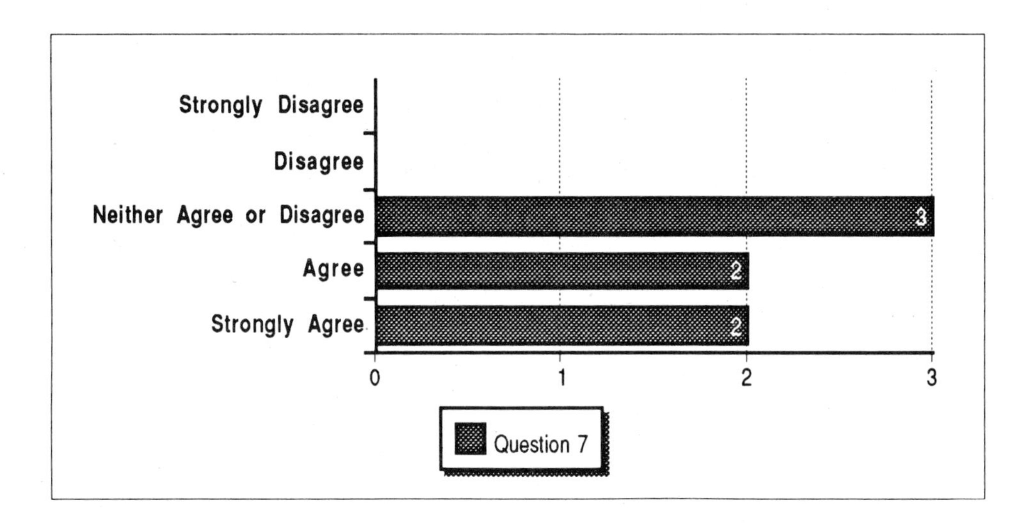

The group recognised the strength of the 'Profiler' and its potential for fast screening. The 'Profiler' is however conceived as a very flexible tool and Q8 attempted to identify whether the participants thought it also good for in-depth studies.

<u>BEFORE WORKSHOP - Part 1: Section 4</u> - Q8 - I need a tool for doing in-depth studies on environmental issues.

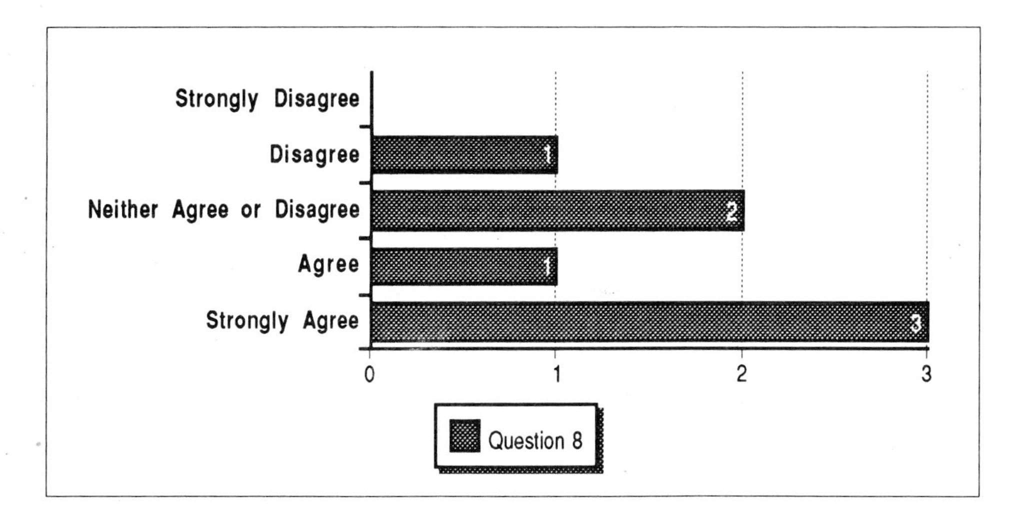

There is a general empathy with the statement.

<u>WORKSHOP END - Part 2: Section 4</u> - *Q8 - Linking one 'Profile' to another will make the 'Profiler' concept useful for in-depth studies.*

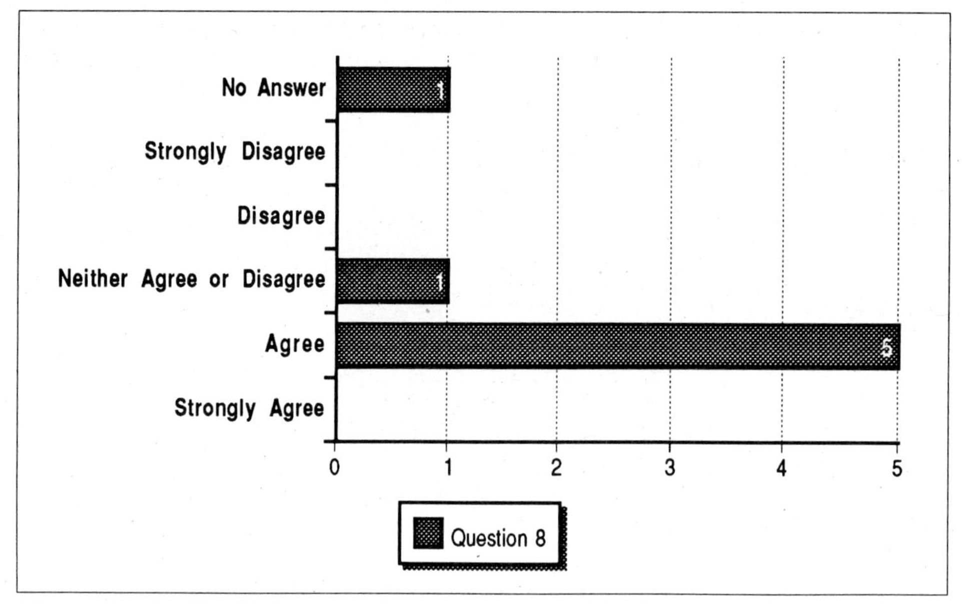

After seeing the 'Profiler' concept the response was positive in support of its potential for in-depth studies. The questionnaire results supported the forecast that the 'Profiler' will be useful for both in-depth and fast screening studies - responses than reflect its flexibility.

4.3 INDUSTRIAL Design Workshop - Coordinator: *Anders Smith*
Nº of Paired Questionnaires: *15 pairs (30 questionnaires in total).*
Sample Questions taken from'Participants Questionnaires: *Part 1: Section 5 (Before workshop)& Part 2: Section 5 (End of workshop)*

<u>BEFORE WORKSHOP - Part 1: Section</u> - *Q13 - Obtaining environmental information is a problem.*

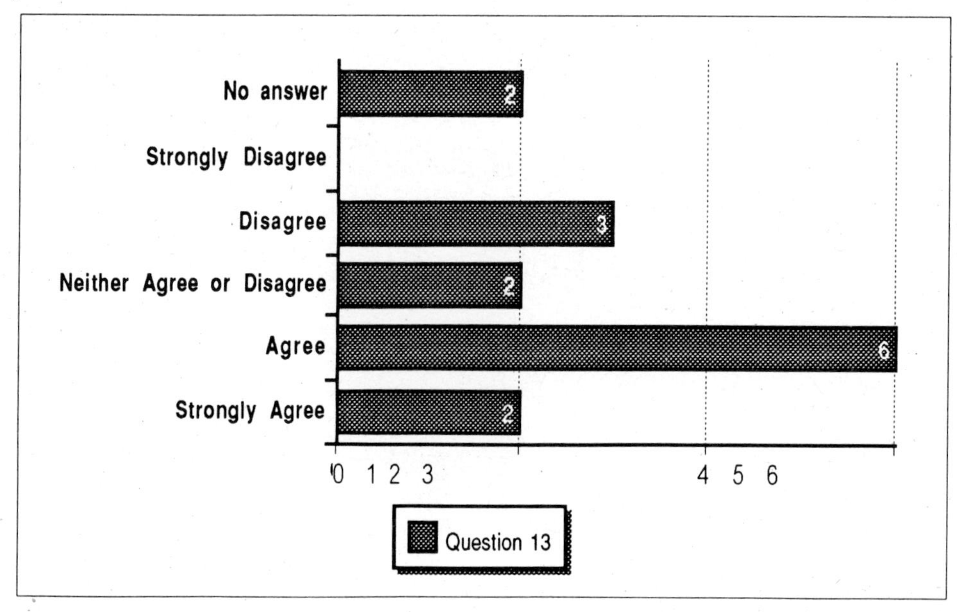

Generally the sentiment of the statement is agreed with. It is important to note that the complexity of obtaining environmental information varies greatly. Considerations are the depth of information required, the breath of the subject to which it is applied, the knowledge of the user, the quality of the information, information availability - commercial secrecy, etc.

WORKSHOP END - Part 2: Section 5 - Q13 - It would be good if the 'Profiler' could obtain information through the 'Internet' in the future.

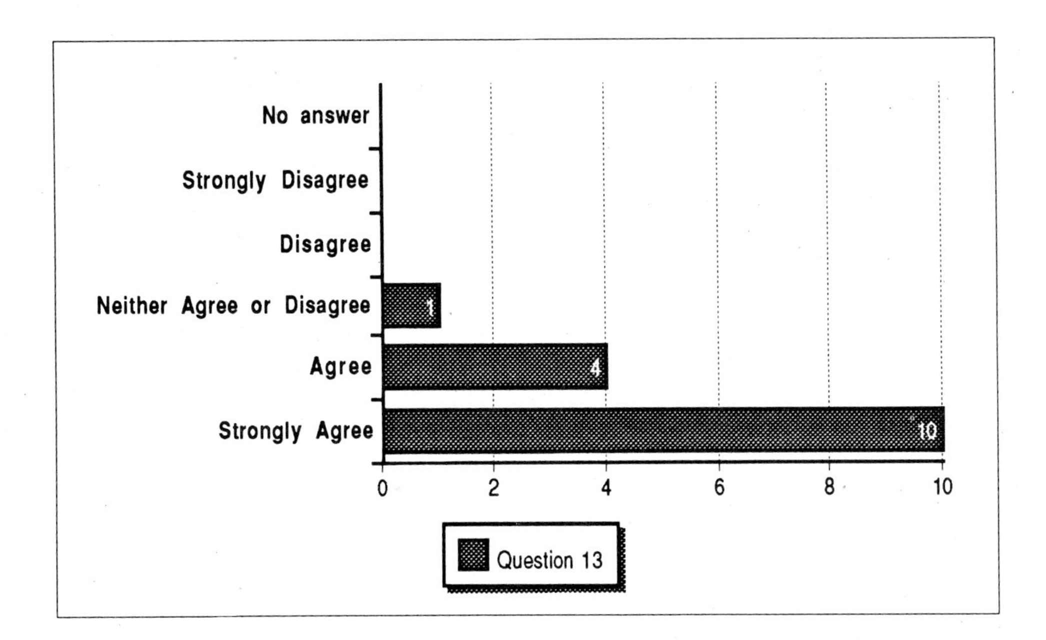

There was an over overwhelmingly positive response to this plan. One participant reinforced this objective with: *'Information is the key to using the 'Profiler'*. Whilst another observed: *'Then arises the problem of data filtering as criticism'*. However, it can be said that data filtering, accessibility and quality of information is a universal problem and not unique to the 'Internet' which is after all only a carrier as is a book or database.

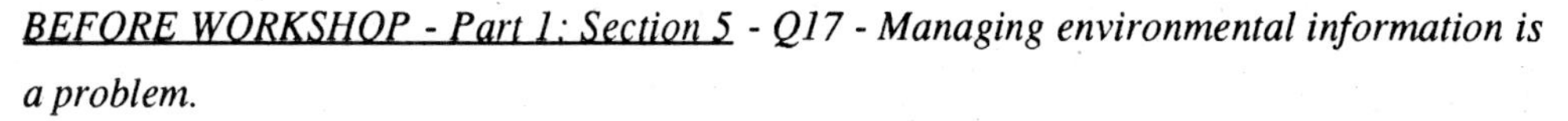
BEFORE WORKSHOP - Part 1: Section 5 - *Q17 - Managing environmental information is a problem.*

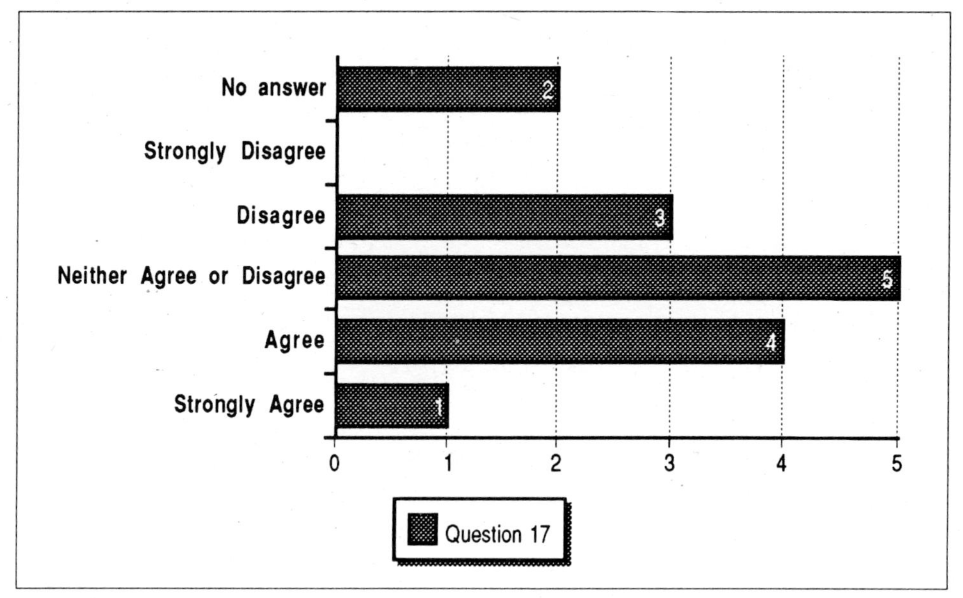

This question gave a more mixed response than had been expected, the result may reflect the different individual experience of the group. Comments varied...*Needs constant updating (information)'... 'Only for inexperienced designers or managers'...'But isn't managing information in design - i.e. information explosion and breath required problematic anyway. Need to be knowledgeable about knowledge'....'There is a variety of software that can be used for this purpose'.*

WORKSHOP END - Part 2: Section 5 - *Q17 - The 'Profiler' concept makes managing environmental information easier.*

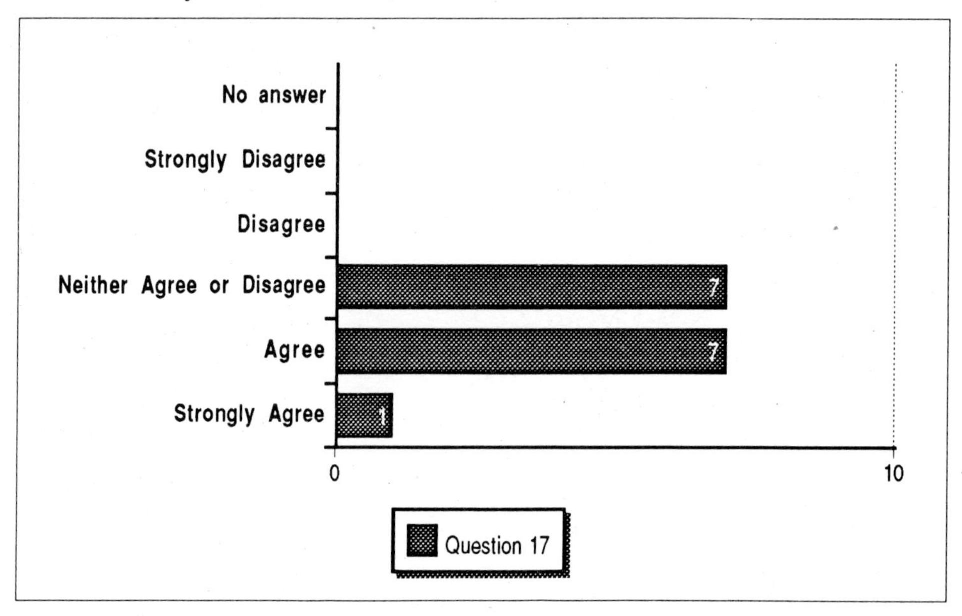

In the main the response was positive towards the statement of Q17. Nonetheless the following comments were made:

'Not sure - easier than what - inability to view the whole picture hides information'...
' It could be useful for interdisciplinary communication. It seems to be a tool able to be transferred from (between) different 'actors' hands'.

On the issue of hiding information it is difficult to imagine how any system could present its entire information (data) input for an aggregated result at one time. The 'Profiler' presents results in two main ways:

1. As a 'graphic' pattern for quick viewing.
2. In a text data file for more in-depth information.

The workshop version allowed 512 single issues to be aggregated which is impossible to show at once. The comment on 'transferring' is very much at the heart of the 'Profiler' design. It is envisaged that users will be able to exchange 'Profiles' to agreed criteria in the future.

4.4 URBAN Planning Workshop - Coordinator: *J.C. van Weenen*

Nº of Paired Questionnaires: *9 pairs (18 questionnaires in total).*

Sample Questions taken from 'Participants Questionnaires: *Part 2: Section 6 (End of workshop). This section focuses on the interface design of the 'Profiler' so there are no matched questions from Part 1. However, an overview of responses to Part 1 Questions already used in the other groups evaluation are shown below.*

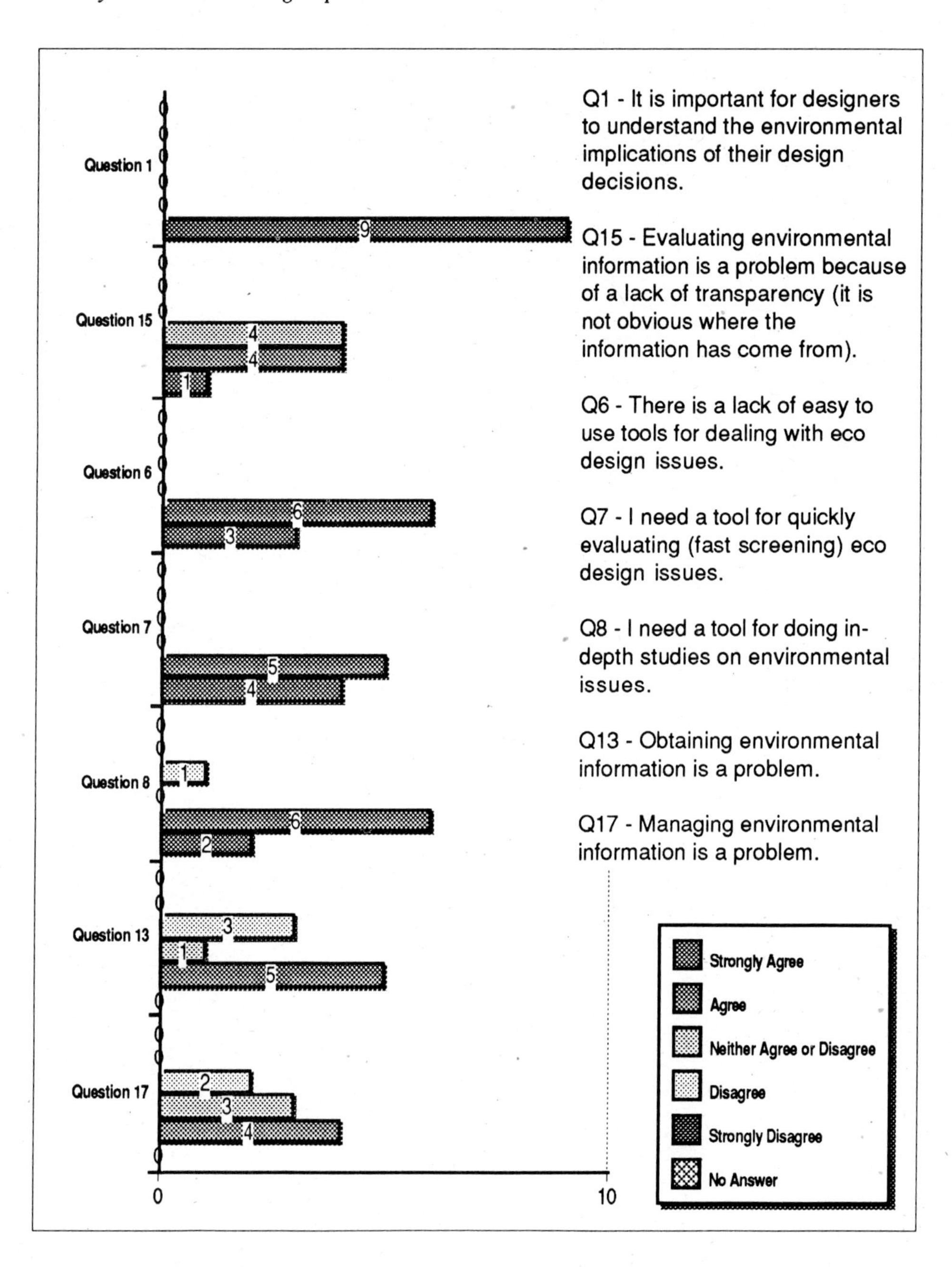

In general the urban planning participants agreed with the question/statements posed. A slight exception being those referring to obtaining and managing information (Q13 & 17).

WORKSHOP END - Part 2: Section 6 - Q18 - Do you find the graphic Profile patterns a useful interface for viewing environmental data.

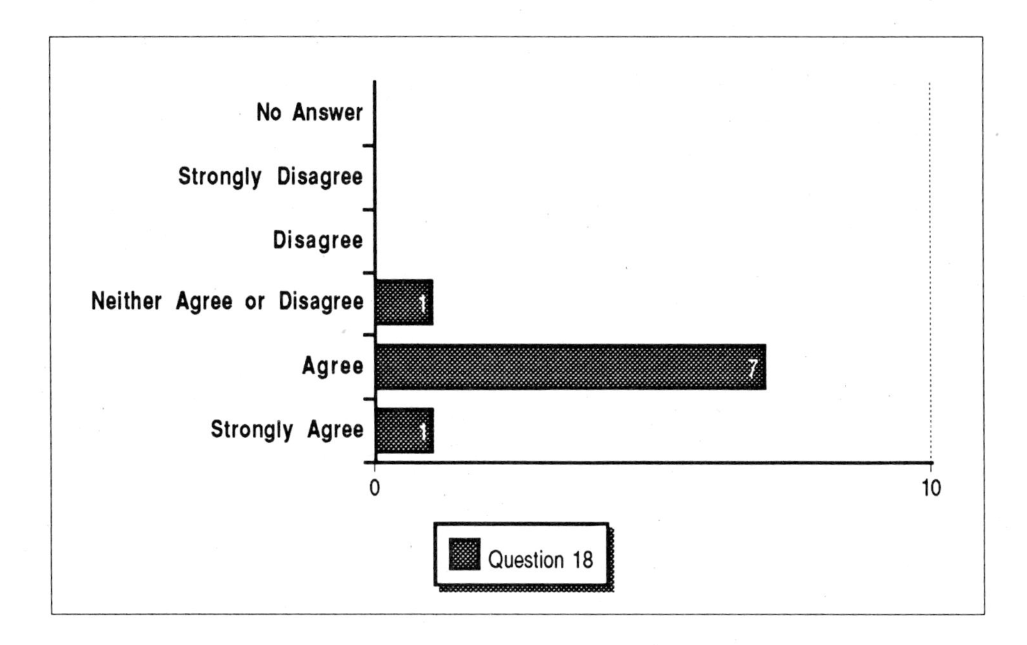

It can be seen that there were very positive replies to this question. However, there were two recommendations on how the 'Profiler' pattern might be improved. One suggested that the pattern area should reduce as environmental performance improves (the 'Profiler' pattern increases in area) the other suggestion was that the lines that made up the pattern suggested that the scores for subjects were influenced by the connections.

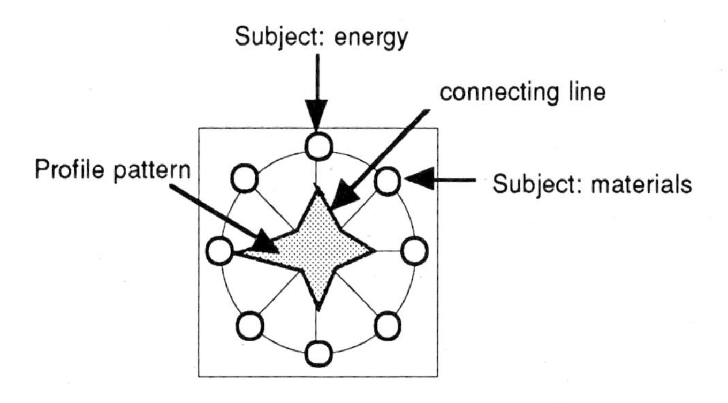

It can be argued that there is a connection as they all contribute to the whole. The 'Profiler' is sub-dividing a whole problem into manageable parts (for example: a product across its whole life cycle) which are environmentally assessed.

With these suggestions in mind the results of all participants Q18 *WORKSHOP END - Part 2: Section 6* can be considered.

WORKSHOP END - Part 2: Section 6 - Q18

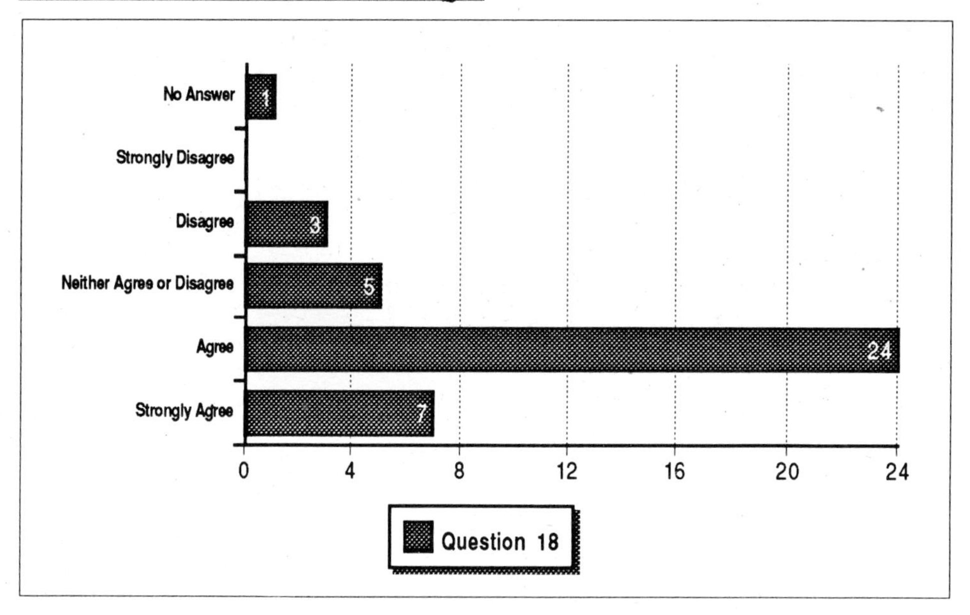

When all results are looked at regarding the 'Profiler' pattern it can be seen that there is very strong support for the approach.

WORKSHOP END - Part 2: Section 6 - Q19 - Do you find the basic 'Profiler' concept simple to understand and accessible in its software form.

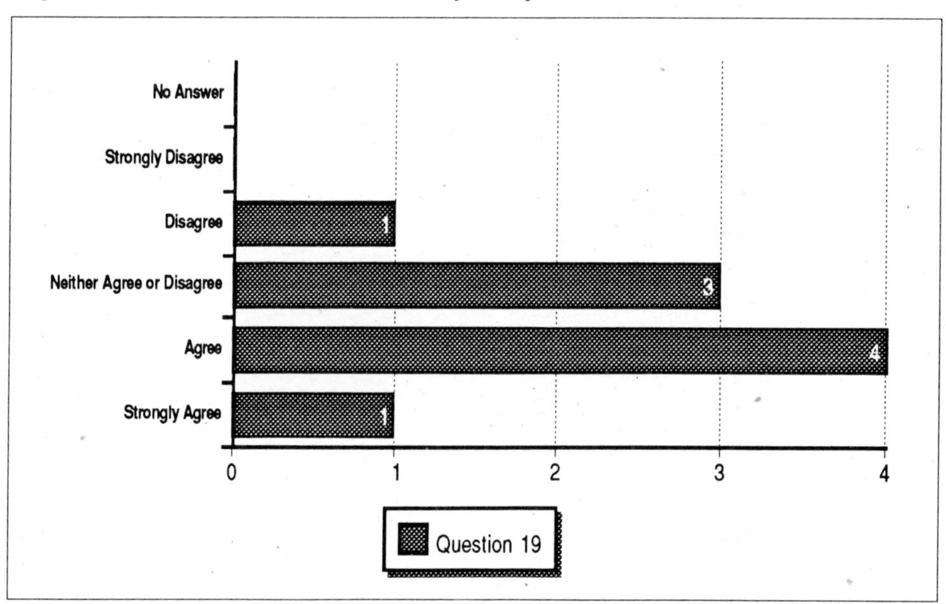

This question poses a very strategic concern and the response is very positive.

5. CONCLUSIONS & RECOMMENDATIONS

Comments from participants' questionnaires are used in the following conclusions drawn from the workshop.

5.1 Conclusions on the 'DfH Profiler'

The overall conclusion was that the majority of participants were overwhelmingly positive about the 'Profiler' approach. This was reflected in the results of Questionnaire Part 2 - Section 7 - Q36. It asked for a response to the statement:

'I would like to use the 'Profiler' in my work.

18.2% Strongly Agreed • 54.5% Agreed • 24.2% Neither Agreed or Disagreed • 3% Disagreed • 0% Strongly Disagreed.

This did not negate the strong opinion and conclusion that the requirement for information gathering and some form of reference criteria must be fulfilled to make the 'Profiler' really useful.

The 'Profiler' is designed to be a very flexible system so that users can develop their own criteria and categories (references) for the situation in which they work. Participants' questionnaires contained the following remarks on the subject of a flexible design...
'Yes - but equally this is a weakness - guidelines and or prompts are needed to guide user' ..

And on the need for references... *'Yes people could see a structure rather than emotive reaction and accept this advice - if there were more references in the programme'...*

Factors inherent in eco design work of criteria setting, using references, information support and measuring were summarised in this remark...*'You need to be able to search for knowledge and guidelines, to evaluate them and weight different aspects against each other'* .

With the above in mind and the result of Q13 below a conclusion is that the 'Profiler' needs to be further developed with reference facilities and information gathering abilities. Questionnaire Part 2 - Section 5 - Q13 - highlighted information gathering:

'It would be good if the 'Profiler' could obtain information through the Internet'

57.1% Strongly Agreed • 34.3% Agreed • 5.7% Neither Agreed or Disagreed • 2.9% Disagreed • 0% Strongly Disagreed.

There was a very clear recognition for the need for a tool to help with product and environment issues, as follows:

' There is a need for adviser type tool' ... 'Environmental issues are of critical importance in our projects- from simplistic specification of 'friendly' materials, to longer term and more complex maintenance, running and use concerns' ... and on the 'Profiler' in particular... *'It will be vital in the future'...* and...*'Well developed argument that needs much wider promotion'....'I see its possible contribution in this field.... 'The 'Profiler' concept would make it easier to advise others on eco design options'.*

Another conclusion is that further studies should be undertaken on behalf of specific user groups and education especially. Many positive remarks were made by those working in education attending the workshops.

...'The positive aspect of the software is its flexibility. I would be interested in using it for educational purposes and introducing the students to environmental considerations in all of their work'....'Easy to understand and use as a method. I see it as a tool for students of design disciplines to make decisions on eco-matters'...

...It should become obligatory for design education in the EU' .

It can also be concluded that the potential for using the 'Profiler' to exchange and share information between different actors was clearly recognised.

...I believe the real benefit is the opportunity of interdisciplinary communication'...'I believe this is an important innovation in the field of Environmental Assessment tools'...'Interesting innovation: Inferences of qualitative and quantitative extremely interesting as interdisciplinary communication tool where 'Life Cycle Design' is an important strategy which requires information exchanges'.

A question that exemplified the need to exchange and explain eco design to different stakes holders was: Questionnaire Part 2 - Section 3 - Q5:*The 'Profiler' concept makes it easier to present eco design options, reasoning & strategies'*

57.1% Strongly Agreed • 34.3% Agreed • 5.7% Neither Agreed or Disagreed • 2.9% Disagreed • 0% Strongly Disagreed.

Recommendations for supporting the conclusions and lessons learnt from the workshops follow.

5.2 Recommendations

The following recommendations would provide valuable research opportunity to evolve a more powerful form of the 'Profiler' whilst assessing its usefulness to different user groups within the EU. It is recommended:

- That further 'exploratory studies' take place to 'evaluate' the Profiler's flexibility in dealing with holistic applications. A suitable subject would be the design of workplaces to improve the quality of working life - which involves specific and holistic concerns of a qualitative and quantitative nature.

- That 'comparison' studies take place between the 'Profiler' and other eco design software tools of both an holistic and specific nature.

- That a study is made on information sources that can be used by the 'Profiler' in guiding a wide range of users on issues of 'sustainable production and consumption'. Specific focus on the World Wide Web.

- That a study is made to 'evaluate' the 'compatibility' of the 'Profiler' with other information sources and users.e.g. how existing information sources can be linked to the 'Profiler' to meet different user needs.

That the 'Profiler' is tested to develop 'case study' experience in the following areas:

To explore consensus building and exchange

Large multinationals to use the 'Profiler' in-house to develop knowledge on the exchange and sharing of information between the different departments, actors and stake holders.e.g. between workers and employers

To explore cost effective guidance and education

SMEs to use the 'Profiler' as a guidance tool by using reference templates from an EU scheme such as EMAS and if possible the more progressive SMAS.

To explore relationships and boundaries

The 'Profiler' is to be used to explore the relationship between the internal and external environments in the context of 'sustainable production and consumption'.

To explore innovative applications

Educational Institutes within the EU to explore areas of application of the 'Profiler'.

The overall focus of the above pilot projects should be the potential of the 'Profiler' to stimulate and support the evolution of more 'sustainable production and consumption' within the EU. And to explore how it can be utilised to exchange, inform and build consensus between the different partners of government, workers and employers.

In support of the software itself it is recommended that:

a. Development of 'Reference' profiles and the ability of the software to overlay one 'Profile' over another is established. Allowing users to measure their project against known targets and criteria.e.g. ISO 14000+.

b. Information wealth of the Internet is made accessible to the 'Profiler' by integrating 'Uniform Resource Locator' (URL) software into the design. Additional information plug in's such as LCA databases and CD-ROM's also need to be supported.

5.3 Conclusions of the Plenary Session on the 'European Foundations' Product and Environment Research. *(By J.C. van Weenen).*

Formulating suggestions and observations from the plenary session into operational actions leads to the presentation of the following action points for the European Foundation:

- The EF should stress in its activities that in auditing for sustainability of both production and consumption, the connection between working conditions and sustainable product development must be addressed. This is e.g. addressed in the 'Sustainable Chemistry' development.

- There is room for an EF-organised European forum and network of developers and designers who are (to be) engaged in sustainable product development.

- The EF should stimulate the realisation of 'Sustainable Regions in Europe', in which industrial activities, building and other processes as well as the products used in buildings are designed for sustainability.

- The EF should support the testing of the 'Profiler' in different types of companies and for specific product groups and products. Practical experience with the 'Profiler' would provide valuable information about its design, practicality and potential.

- An assessment should be made on behalf of the EF, on how companies can realise the management of sustainable product development.

- That the European Foundation provides the platform in which the social partners should address the issue of sustainable product design in a social dialogue. In conjunction with this, an 'Advisory Board' of designers should be formed with the task to create and formulate a programme for the stimulation of sustainable product development.

- The EF should contribute to the realisation of an acknowledged registration system and an organisation for designers who have studied and practised eco-design, who have specialised and have sufficient experience.

- The EF should put in place incentives such as award schemes which stimulate Sustainable Product Development, in education as well as in industrial practice.

6. REFERENCES

'New Materials for Environmental Design', European Foundation for the Improvement of Living & Working Conditions (1994), ISBN 92-826-8612-4.

The above is available in French, European Foundation for the Improvement of Living & Working Conditions (1995): ISBN 92-826-8613-2.

Benjamin. Y. 'The Design for Health Eco Profiler, Information Management in a Computing Environment', European Foundation for the Improvement of Living & Working Conditions, (1995), Working paper: WP/95/09/EN.

Design for Health - Creating a Brochure, European Foundation for the Improvement of Living & Working Conditions (1995), ISBN 92-827-4114-1.

7. ANNEX 1
Participants Questionnaire Part 1 (sent prior to workshop)

Please note: Questions started in the original document on Section 3
ALL MATCHED QUESTIONS ARE IN SECTIONS 3,4,5

3.0 ECO DESIGN QUESTIONS - OVERVIEW

1. It is important for designers to be able to understand the environmental implications of their design decisions.
2. I would like to do more eco design.
3. Clients (or people I work with) expect me to be competent in this area.
4. It is difficult to advise others (e.g. clients) on environmental issues.
5. It is difficult to present eco design findings so that they are understandable by others.

4.0 ABOUT ECO DESIGN TOOLS

6. There is a lack of easy to use tools for dealing with eco design issues.
7. I need a tool for quickly evaluating (fast screening) eco design issues.
8. I need a tool for doing in depth studies on environmental impacts.
9. Eco design tools in the form of handbooks are too general & date quickly.
10. I have found that eco design tools in the form of software (e.g. LCA database) complex and unsympathetic to the non specialist.
11. Eco design tools do not facilitate qualitative design concerns.e.g. colour, emotion, fashion, cultural & social issues, etc.
12. It is important that qualitative design concerns (re: 11) are included in eco design studies as these issues can greatly influence a products impact.

5.0 ABOUT ENVIRONMENTAL INFORMATION

13. Obtaining environmental information is a problem.
14. Understanding environmental data/information is a problem.
15. Evaluating environmental information is a problem because of a lack of transparency (it is not obvious where the information has come from).
16. Comparing (weighting) environmental information is a problem.
17. Managing environmental information is a problem.

6.0 ABOUT LCA

18. LCA methods are not useful for design work.
19. LCA results are not easy to understand.
20. LCA results are difficult for 'clients' to understand.
21. LCA results are difficult to communicate to others.
22. LCA is generally too expensive.
23. LCA studies are too removed from the design process to be useful.

ANNEX 2

Participants Questionnaire Part 2 (completed at end of workshop)

Please note: Questions started in the original document on Section 3
ALL MATCHED QUESTIONS ARE IN SECTIONS 3,4,5

3.0 'PROFILER' CONCEPT - OVERVIEW

1. Using the 'Profiler' concept would help me to understand the environmental implications of my design decisions.
2. The 'Profiler' concept would help me to consider environment and health impacts more often in my work.
3. The 'Profiler' concept would help me to consider environment and health impacts more often in my work.

4&5. The 'Profiler' concept makes it easier to advise others on eco design options and to present them.

4.0 ABOUT THE 'PROFILER' CONCEPT

6. The 'Profiler' concept was easy to use.
7. The 'Profiler' concept is good for fast screening environmental issues.
8. Linking one 'profile' to another will make the 'Profiler' concept useful for in depth studies.
9. The 'Profiler' concept can be applied as the user demands (e.g. it can meet the specific project needs of users).
10. The 'Profiler' concept is easy enough to be accessible to the non specialist.
11. The 'Profiler' concept is very useful at managing and applying values to qualitative design issues. e.g. colour, emotion, fashion, cultural & social issues, etc.
12. The 'Profiler' concept would enable me to incorporate qualitative design values in environmental assessments.

5.0 'PROFILER' & INFORMATION

13. It would be good if the 'Profiler' could obtain information through the 'Internet' in the future.
14. The way the 'Profiler' concept presents environmental information makes it understandable.
15. The way the 'Profiler' concept develops its environmental 'profile' makes the information very transparent.
16. The 'Profiler' concept makes it easy to compare environmental information.
17. The 'Profiler' concept makes managing environment information easier.

6.0 'PROFILER' INTERFACE DESIGN

18. Do you find graphic profile patterns a useful interface for viewing environmental data.
19. Do you find the basic 'Profiler' concept simple to understand and accessible?
20. It was easy to create a new profile.
21. It was easy to create a new topic spoke.
22. It was easy to change the position of a topic spoke.
23. There are insufficient topic spokes.
24. There are insufficient numbers of profiles.
25. It was easy to access and change a topic title.

26. The topic title string is too short.
27. It was easy to adjust the value of a topic spoke.
28. It was easy to change the information type (bias) of a topic spoke.
29. It was easy to find and access other profiles.
30. It was easy to link profiles together.
31. It was easy to access the note facility.
32. It was easy to create and organise comments in the notes facility.
33. The menus and controls are located logically.
34. The menus and controls are easy to use.
35. The overall interface was simple (as opposed to complex).

7.0 'PROFILER' APPLICATIONS

36. I would like to be able to apply the 'Profiler' in my work.

And a tick box response on:

If you used the 'Profiler' - how would you use it in your work?

☐ Fast screening of existing products.
☐ For evaluating concepts.
☐ Developing quantitative Life Cycle Assessment (LCA)
☐ For inventory data. e.g. to make an environmental audit of an existing product, package or process.
☐ For environmental management systems (EMS)
☐ For developing policy.
☐ Other.

ANNEX 3 • LIST OF PARTICPANTS

A multidisciplinary group of designers, social partners and potential users of the software were invited.

Graphic Design

Chair : Helmut Langer

Mr. Helmut LANGER
Icograda
Postfach 510713
D-5943 Köln
GERMANY
Tel + 49 221 388 729
Fax + 49 221 342 985

Ms. Anne CHICK
Faculty of Design
Falkner Road
Farnham
Surrey GU 9 7DS
UNITED KINGDOM
Tel + 44 1 252 732 229
Fax + 44 1 252 732 274

Ms. Karen BLINCOE
Head of Institute for Visual Communication
Danmarks Designskole
Strandboulevarden 47
DK-2100 Copenhagen
DENMARK
Tel + 45 35 27 75 19
Fax + 45 35 27 76 00

Ms.Mary-Ann LINDHOLM
c/o J.W. Tompson
Bulevardi 42
SF-00120 Helsinki
FINLAND
Tel + 358 0 618 8345
Fax + 358 0 618 83499

Mr. Anders SUNESON
Graphic Designer
The Swedish Association of Illustrators and Graphic Designers
P1 1802 Härke
S-832 00 Frösön
SWEDEN
Tel + 46 63 44 153
Fax + 46 63 43 296

Ms. Angela BALDINGER
Zahnradbahnstr 5
A-1190 Vienna
AUSTRIA
Tel + 43 1 375 777
Fax + 43 1 375 455

Mr. Yiannis KOUROUDIS
Graphic Designer
104, Lavriou St.
Alsoupolis 14235
Athens
GREECE
Tel + 30 1 275 4245
Fax + 30 1 275 4245

Mr. Mark STRACHAN
19 Kingsdown Avenue
Ealing
London W13 9PS
UNITED KINGDOM
Tel + 44 181 567 2702
Fax + 44 181 567 0704

Ms. Anna BENTZEN
Project Manager
Kontrapunkt
Christians Brygge 28
DK-1559 Copenhagen V
DENMARK
Tel + 45 33 93 1883
Fax + 45 33 93 1854

Mr. Philip IRELAND
Publications Assistant
European Foundation
Loughlinstown House
Shankill, Co. Dublin
IRELAND
Tel + 353 1 282 6888
Fax + 353 1 282 6456

Ms. Kaarina POHTO
Editor in Chief
ICSID Secretariat
Yrjönkatu 11 E20
FIN-00120 Helsinki
FINLAND
Tel + 358 0 607 611
Fax + 358 0 607 875

Mr. Alfredo DE SANTIS
Graphic Designer
Via Pontedera 9
I-00161 Rome
ITALY
Tel + 39 6 4424 5464
Fax + 39 6 4424 5121

Mr. Ilmo VALTONEN
Graphic Designer
Indigo Oy and Grafiary
Association of Finnish Graphic Designers
Bulevardi 6
FIN-00120 Helsinki
FINLAND
Tel + 358 0 125 5475
Fax + 358 0 641 743

Furniture Design
Chair: Yorick Benjamin

Mr. Yorick BENJAMIN
EDEN
Dewittenkade 93hs
1052 AE Amsterdam
THE NETHERLANDS
Tel + 31 20 488 4251
Fax + 31 20 625 88 43

Mr. Denis HANDY
President
Ryan O'Brien Handy
6 Percy Place
Dublin
IRELAND
Tel + 353 1 668 0899
Fax + 353 1 668 0089

Mr. Richard LINIGTON
The Chadwich Group
1A Birkenhead Street
London WC1 8NB
UNITED KINGDOM
Tel + 44 171 278 5969
Fax + 44 171 833 1621

Mr. Howard LIDDELL
Principal Architect
Gaia Architects
Aberfeldy Studios
Chapel Street
Perthshire PH15 2AW
UNITED KINGDOM
Tel + 44 1 887 820 160
Fax + 44 1 887 829 544

Ms. Marianne FRANDSEN
Interior Architect MDD
Den Blaa Tegnestue
Gothersgade 8 B
1123 Copenhagen K
DENMARK
Tel + 45 33 14 9104
Fax + 45 33 933 004

Ms. Efi VALIANTZA
Environmental Engineering Consultant
FIVI, Cleaner Production Center for Greek SME's
National Research Center 'Demokritos', Ag Paraskevi
Athens 15310
GREECE
Tel + 30 1 654 9479
Fax + 30 1 653 6531

Mr. Carlo PESSO
Product Policy Analyst
OECD
2, Rue Andre Pascal
75775 Paris Cedex
FRANCE
Tel + 33 1 45 24 1682
Fax + 33 1 45 24 7876

Mr. Jacob FALKENBERG
Irish Productivity Centre
IPC House
42-47 Lower Mount St.
Dublin 2
IRELAND
Tel + 353 1 66 23 233
Fax + 353 1 66 23 300

Ms. Camille DERMONNE
Eco-Counseil
Institut Eco-Conseil
rue Galliot, 20
B-5000 Namur
BELGIUM
Tel + 32 81 24 1101
Fax + 32 81 23 1681

Mr. Thierry VAN KERM
Managing Director
NAOS Design
85 rue des Glands
B-1190 Brussels
BELGIUM
Tel + n/a
Fax + n/a

Ms. Ann MARINELLI
Sole Design
Corso C. Colombo 9
20144 Milan
ITALY
Tel + 39 2 894 00 172
Fax + 39 2 894 00 172

Mr. Vilhelm Lange LARSSEN
Advisor
Norwegian Design Council
Riddervoldsgt. 2
N-0256 Oslo
NORWAY
Tel + 47 2 558 040
Fax + 47 2 559 302

Mr. Carlo VEZZOLI
Management and Production Engineer
Research Unit
Polytechnic University of Milan
Via Bonardi 3
20133 Milan
ITALY
Tel + 39 2 2399 5124
Fax + 39 2 2399 5130

Mr. Jim HALPENNY
Computer Services
European Foundation
Loughlinstown House
Shankill
Co. Dublin
IRELAND
Tel + 353 1 282 6888
Fax + 353 1 282 6456

Mr. Bob VERHEIJDEN
Head of 3D Department
Hogeschool voor de Kunsten Arnhem
Onderlangs 9
6812 CE Arnhem
THE NETHERLANDS
Tel + 31 85 535 621
Fax + 31 85 535 677

Mr. Anders KARDBORN
Institutionen för
Konsumentteknik
Chalmers University of
Technology
S-41296 Götegborg
SWEDEN
Tel + 00 46 31 772 3598
Fax + 00 46 31 772 1111

Mr. Elfy MEYER
Atelier d_Art Graphique
13 Avenue du Souleilla
F-31320 Vigoulet-Auzil
FRANCE
Tel/Fax + 33 61 733 390

Industrial Design
Chair: Anders Smith

Mr. Anders SMITH
Managing Director
Kontrapunkt A/S
Christians Brygge 28
DK-1559 Copenhagen V
DENMARK
Tel + 45 33 93 18 83
Fax + 45 33 93 18 54

Mr. H.C. HOLMSTRAND
Senior Environment
Consultant
ECONET AS
Vesterbrogade 26
DK-11620 Copenhagen
DENMARK
Tel + 45 31 24 6522
Fax + 45 31 24 6518

Mr. Stefan PANNENBECKER
Product Designer
Philips Corporate Design
Cederlaan 4
5616 SC Eindhoven
THE NETHERLANDS
Tel + 31 40 732 510
Fax + 31 40 733 482

Mr. John BARRETT
Product Designer
Philips Corporate Design
Building OAN
5616 SC Eindhoven
THE NETHERLANDS
Tel + 31 40 732 671
Fax + 31 40 733 482

Dr. Gerard ZWETSLOOT
Netherlands Institute for the
Working Environment
(NIA)
P.O. Box 75665
THE NETHERLANDS
Tel + 31 20 549 8449
Fax + 31 20 549 8530

Mr. David ROTHWELL
10 Melbourne Avenue
Drumcondra
Dublin
IRELAND
Tel + 353 1 83 78 955

Dr. Mohan EDIRISINGHE
Department of Materials
Technology
Brunel University
Uxbridge
Middlesex UB8 3PH
UNITED KINGDOM
Tel + 44 1895 274 000 ext
2981
Fax + 44 1895 812 636

Prof. BONSIEPE
Hypermedien und Interface
Design
FHS FB Design
Ubierrung 40
D-50678 Köln
GERMANY
Tel + 49 221 8275 3200
Fax + 49 221 3188 22

Ms. PAAKKUNAINEN
University of Art and Design
Department of Textile Art
and Fashion Design
Hämeentie 135 C
FIN-00560 Helsinki
FINLAND
Tel + 358 0 756 30 333
Fax + 358 0 756 30 223

Ms. Ursula TISCHNER
Diplom Designer
Wuppertal Institut
Döppersberg 19
D-42103 Wuppertal
GERMANY
Tel + 49 202 2492 163
Fax + 49 202 2492 138

Mr. Roland WHITEHEAD
19 Kingsdown Avenue
Ealing
London W13 9PS
UNITED KINGDOM
Tel + 44 181 567 2702
Fax + 44 181 567 0704

Mr. Peik SUYLING
Hahn & Suyling Design
Lindenstraat 5
1015 KV Amsterdam
THE NETHERLANDS
Tel + 31 20 623 5181
Fax + 31 20 639 1892

Mr. Phillip GOGGIN
Eco Design Coordinator
Design Studies
Goldsmiths College
University of London
New Cross,
London SE14 6NW
UNITED KINGDOM
Tel + 44 171 919 7756
Fax + 44 171 919 7786

Prof. RAHE
Industrial Designer
Am WeißDFenhof 21
D-70191 Stuttgart
GERMANY

Urban Planning and Sustainability
Chair: Hans van Weenen

Dr. Hans VAN WEENEN
UNEP
University of Amsterdam
J.H. van_T Hoff Institute
Nieuwe Achtergracht 166
1018 WV Amsterdam
THE NETHERLANDS
Tel + 31 20 525 5859
Fax + 31 20 625 8843

Ms. Anne-Michéle DONNET
2 bis, rue de l'Arrivée
92190 Meudon
FRANCE
Tel + 33 1 4534 7497
Fax + 33 1 4626 4245

Mr. Armando MONTANARI
ISEMEM-CNR
Via A Gramsci, 5
I-80122 Naples
ITALY
Tel + 39 81 664 553
Fax + 39 81 761 3993

Mr. Antoni RAMON
Centro Internacional de Estudio Urbanos (CIEU)
Calabria 13-15
08015 Barcelona
SPAIN
Tel + 34 3 425 2422
Fax + 34 3 426 6213

Ms. Gabriele TAGLIAVENTI
Architect
Vision of Europe
University of Bologna
Viale Risorgimento, 2
Bologna
ITALY
Tel + 39 51 644 3160 / 233 717
Fax + 39 51 222 329

Mr. Stavros TSETSIS
Architect Planner Engineer
Stavros Tsetsis & Associates
Stratiotikou Syndesmou 4
GR-106 73 Athens
GREECE
Tel + 30 1 362 5070
Fax + 30 1 362 5070

Mr. Stephen PURVIS
Charterhouse Regeneration Ltd,
33A Talbot Road
London W2 5JG
UNITED KINGDOM
Tel + 44 171 229 5700
Fax + 44 171 229 1912

Ms. Carmen SERRANO
D.G. Politica Ambiental (MOPTMA)
Despacho A-427
P°Castellana, 67
28071 Madrid, SPAIN
Tel + 34 1 59 77 480
Fax + 34 1 59 78 513

Professor J.F. OLIVEIRA
Full Professor
Univ Nova Lisboa
GDEH/FCT/UNL
Quinta Da Torre
2825 Monte da Caparica
PORTUGAL
Tel + 351 1 295 4464
Fax + 351 1 295 4461

Mr. Garsett LAROSSE
Director
Ecotopia Sustainability Center
Antwerpsesteenweg 461
B-2390 Westmalle
BELGIUM
Tel + 32 3 309 2873
Fax + 32 3 309 2874

Dr. Ekhart HAHN
Gesellschaft für Ökologischen Städtebau und Stadtforschung
Paul-Lincke-Ufer 30
1000 Berlin 36
GERMANY
Tel + 49 30 611 8511
Fax + 49 30 618 9005

Mrs. Anna Marie PETERSEN
Architect MAA
Bro/Petersen Architects MAA
Skorup Gl. Skole
DK-8882 Fårvang
DENMARK
Tel + 45 868 710 15
Fax + 45 868 722 15

Ms. Eva ÖRNEBLAD
Post Graduate Student
Chalmers University of Technology
Industrial Architecture and Planning
412 96 Gothenburg
SWEDEN
Tel + 46 31 772 2337
Fax + 46 31 772 2461

Mr. Per BRO
Architect
Skorup Gamle Skole
DK-8882 Fårvang
DENMARK
Tel + 45 8687 1015
Fax + 45 8687 2215

Mr. Ole GEISLER
Consultant
Dansk Miljøanalyse Aps
Science Park Symbion
Fruebjergvej 3
DK-2100 Copenhagen Ø
DENMARK
Tel + 45 39 17 99 22
Fax + 45 31 20 68 89

Participants who freely choose workshop according to interests follow.

Trade Unions

Mr. Michel MILLER
Assistant
European Trade Union Confederation,
Boulevard Emile Jacqmain 155,
B-1210 Brussels
BELGIUM
Tel + 32 2 224 0411
Fax + 32 2 224 0454/5

Employers

Mr. Bernard LE MARCHAND
Coordinateur Adjoint du Groups des Employeurs
FEMGED
Avenue Victor Gilsoul 76
1200 Brussels
BELGIUM
Tel + 32 2 771 5871
Fax + 32 2 771 5871

Governments

Dr. Uwe WÖLCKE
Head of "Hazardous substances Assessment Authority under the Chemical Act"
Federal Institute for Occupational Safety & Health
Postfach 170202
D-44061 Dortmund
GERMANY
Tel + 49 231 907 1323
Fax + 49 231 907 1679

European Foundation

Mr. Eric VERBORGH
Deputy Director
European Foundation
Loughlinstown House
Shankill
Dublin
IRELAND
Tel: + 353 1 282 6888
Fax: + 353 1 282 6456

Mr. Henrik LITSKE
Research Manager
European Foundation
Loughlinstown
Dublin 18
IRELAND
Tel: + 353 1 282 6888
Fax: + 353 1 282 6456

Others

Mrs. Birgit CRAMON-DAIBER
General Secretary
Focus Europe
Elßholzstrasse 7
D-10781 Berlin
GERMANY
Tel + 49 30 216 5783
Fax + 49 30 215 3779

Mr. Jorn E. BEHAGE
Designer
Kiem Product Development Support
Leliegracht 19
1016 GR Amsterdam
THE NETHERLANDS
Tel + 31 20 638 5678
Fax + 31 20 638 5678

Mr. John ULHØI
Assistant Professor
The Aarhus School of Business
Department of Organisation and Management
Haslegårdsvej 10
DK-8210 Århus
DENMARK
Tel + 45 89 48 6688
Fax + 45 86 15 7629

European Foundation for the Improvement of Living and Working Conditions

European Workshops on Eco Products
– Proceedings

Luxembourg: Office for Official Publications of the European Communities

1996 – 60 pp. – 21cm x 29.7 cm

ISBN 92-827-7790-1

Price (excluding VAT) in Luxembourg: ECU 7